AVERAGES:

A New Approach

*

Jane Grossman
University of Lowell

Michael Grossman
University of Lowell

Robert Katz
Archimedes Foundation

1983

Archimedes Foundation
Box 240, Rockport
Massachusetts 01966

ISBN 0977117049

First Printing, 1983

PREFACE

This monograph is primarily concerned with a comprehensive family of averages of functions that arose naturally in our development of non-Newtonian calculus [2] and weighted non-Newtonian calculus [5]. The approach is based on the idea that each ordered pair of arithmetics (slightly specialized complete-ordered-fields) determines various averages (unweighted and weighted) in a simple manner that we shall explain fully. The monograph also contains discussions of some heuristic guides for the appropriate use of averages and an interesting family of means of two positive numbers.

In Chapter 1 we discuss arithmetic averages in a manner that leads naturally to the other averages presented later. Chapter 2 has a discussion of geometric averages that parallels the presentation of arithmetic averages in Chapter 1.

Chapter 3 contains a discussion of nonclassical arithmetics, which are arithmetics distinct from classical arithmetic (the real number system). In each nonclassical arithmetic, as well as in classical arithmetic, there is a "natural" method for averaging numbers. (For example, in geometric arithmetic the natural average is the geometric average; in classical arithmetic the natural average is the arithmetic average.) Apparently, prior to the creation of non-Newtonian calculus in 1967, no one had conceived the idea of using nonclassical arithmetics for the construction of averages of functions or for any other purpose, it having been long believed that there is no distinctive value in the nonclassical arithmetics, since they are all structurally equivalent to classical arithmetic. (Nonclassical arithmetic should be distinguished from the nonstandard arithmetic created by Abraham Robinson.)

In Chapter 4 we present a general theory of averages of functions, we indicate the uniform relationship between the generalized averages and the arithmetic averages, and we discuss some specific averages of special interest.

Chapter 5 contains a variety of heuristic principles for making appropriate choices of averages. In Chapter 6, we use various averages of the identity function to construct an infinite family of means of two positive numbers.

Various digressions and comments have been placed in the NOTES at the ends of the sections. An annotated bibliography, a list of symbols, and an index have been provided at the end of the monograph.

Since this self-contained work is intended for a wide audience, including students, engineers, scientists, and mathematicians, we have included many details that would not appear in a research report, and we have excluded proofs, most of which are straightforward.

Suggestions and criticisms are invited.

Jane Grossman
Michael Grossman
Robert Katz

CONTENTS

PRELIMINARIES

The word number means real number. The letter R stands for the set of all numbers.

If $r < s$, then the interval $[r,s]$ is the set of all numbers x such that $r \leq x \leq s$. (Only such intervals are used here.)

An arithmetic partition of $[r,s]$ is any arithmetic progression whose first and last terms are r and s, respectively. An arithmetic partition with exactly n terms is said to be n-fold.

A point is any ordered pair of numbers, each of which is called a coordinate of the point. A function is any set of points, each distinct two of which have distinct first coordinates.

The domain of a function is the set of all its arguments (first coordinates); the range of a function is the set of all its values (second coordinates). A function is said to be on its domain, defined at each of its arguments, onto its range, and into each set that contains (is a superset of) its range.

A positive function is any function whose values are all positive numbers.

If every two distinct points of a function f have distinct second coordinates, then f is one-to-one and its inverse, f^{-1}, is the one-to-one function consisting of all points (y,x) for which (x,y) is a point of f.

The identity function is the function I on R such that

$I(x) = x$ for each number x.

The function exp is on R and assigns to each number x the number e^x, where e is the base of the natural logarithm function, ln. The function ln is the inverse of exp.

Finally, for our purposes, it is convenient to use the symbol $\int_r^s f$, rather than $\int_r^s f(x)\,dx$, to represent the 'classical' integral of a continuous function f on [r,s]. Of course, $\int_r^r f = 0$ and $\int_s^r f = -\int_r^s f$.

CHAPTER ONE

ARITHMETIC AVERAGES

1.1 INTRODUCTION

Of the infinitely-many averages of functions, the unweighted and weighted arithmetic averages are the best known and most widely used. Arithmetic averages are easy to calculate, and they provide powerful and intuitively-satisfying insight into many scientific ideas. (We provide an example in Section 5.2.)

In this chapter we shall discuss arithmetic averages in a manner that leads naturally to the other averages presented in subsequent chapters.

1.2 ARITHMETIC AVERAGE

Our definition of the arithmetic average of a continuous function on an interval is based on arithmetic partitions and the following familiar concept.

The <u>arithmetic average of n numbers</u> $z_1, \ldots, z_n$ is the number $(z_1 + \cdots + z_n)/n$.

The <u>arithmetic average of a continuous function f on an interval [r,s]</u> is denoted by $M_r^s f$ and is defined to be the limit of the convergent sequence whose nth term is the arithmetic average

$$[f(a_1) + \cdots + f(a_n)]/n,$$

where $a_1,\ldots,a_n$ is the n-fold arithmetic partition of $[r,s]$.[1] We set $M_s^r f = M_r^s f$.

It turns out that

$$M_r^s f = \frac{1}{s - r} \int_r^s f .$$

The operator M is additive, subtractive, and homogeneous; that is, if f and g are continuous on $[r,s]$, then

$$M_r^s(f + g) = M_r^s f + M_r^s g,$$
$$M_r^s(f - g) = M_r^s f - M_r^s g,$$
$$M_r^s(c \cdot f) = c \cdot M_r^s f, \quad \text{c any constant.}$$

The operator M is characterized by the following three properties. (This use of the word "characterized" indicates that no other operator possesses all three properties.)

For any interval $[r,s]$ and any constant function $k(x) = c$ on $[r,s]$,

$$M_r^s k = c.$$

For any interval $[r,s]$ and any continuous functions f and g, if $f(x) \leq g(x)$ on $[r,s]$, then

$$M_r^s f \leq M_r^s g.$$

For any numbers r, s, t such that $r < s < t$ and any continuous function f on $[r,t]$,

$$(s - r) \cdot M_r^s f + (t - s) \cdot M_s^t f = (t - r) \cdot M_r^t f.$$

We conclude this section with the following mean value theorem:

If f is continuous on $[r,s]$, then strictly between

r and s there is a number c such that

$$M_r^s f = f(c);$$

that is, the arithmetic average of a continuous function on [r,s] is assumed at some argument strictly between r and s.

NOTE

1. The arithmetic average fits naturally into the classical calculus of Newton and Leibniz in a manner explained fully in [2].

1.3 CLASSICAL WEIGHT FUNCTIONS

Our discussion of weighted arithmetic averages in Section 1.5 will be facilitated by the definitions and results in this section and the next.

A classical weight function is any continuous positive function on R. There are, of course, infinitely-many classical weight functions.

For the remainder of this chapter w is an arbitrarily selected classical weight function.

1.4 CLASSICAL MEASURES

The <u>classical measure of an interval [r,s]</u> is the positive number s - r.

Now observe that the classical measure of [r,s] equals the limit of the constant (and hence convergent) sequence whose nth term is the sum

$$1 \cdot (a_2 - a_1) + \cdots + 1 \cdot (a_n - a_{n-1}),$$

where $a_1,\ldots,a_n$ is the n-fold arithmetic partition of [r,s]. This suggests the following definition.

The <u>ω-classical-measure of an interval [r,s]</u> is the positive limit of the convergent sequence whose nth term is the weighted sum

$$\omega(a_1) \cdot (a_2 - a_1) + \cdots + \omega(a_{n-1}) \cdot (a_n - a_{n-1}),$$

where $a_1,\ldots,a_n$ is the n-fold arithmetic partition of [r,s].

We shall use the symbol m[r,s] for the ω-classical-measure of [r,s].

Clearly

$$m[r,s] = \int_r^s \omega .$$

Of course, if $\omega(x) = 1$ on [r,s], then m[r,s] equals s - r, the classical measure of [r,s].

Furthermore, if $r < t < s$, then

$$m[r,t] + m[t,s] = m[r,s].$$

1.5 WEIGHTED ARITHMETIC AVERAGES

We remind the reader that w is an arbitrarily selected classical weight function.

Recall that our definition of the arithmetic average of a continuous function on an interval is based on arithmetic partitions and unweighted arithmetic averages (of n numbers). Our definition of the w-arithmetic average of a continuous function on an interval is based on arithmetic partitions and weighted arithmetic averages (of n numbers), which we define next.

Let $v_1, \ldots, v_n$ be any n positive numbers. For any n numbers $z_1, \ldots, z_n$, the number

$$\frac{v_1 z_1 + \cdots + v_n z_n}{v_1 + \cdots + v_n}$$

is called a <u>weighted arithmetic average</u> of $z_1, \ldots, z_n$.

The <u>w-arithmetic-average of a continuous function f on an interval [r,s]</u> is denoted by $\underline{\underline{M}}_r^s f$ and is defined to be the limit of the convergent sequence whose nth term is the weighted arithmetic average

$$\frac{w(a_1) \cdot f(a_1) + \cdots + w(a_n) \cdot f(a_n)}{w(a_1) + \cdots + w(a_n)},$$

where $a_1, \ldots, a_n$ is the n-fold arithmetic partition of [r,s].[1]

It turns out that $\underline{\underline{M}}_r^s f$ equals the well-known weighted arithmetic average of f on [r,s]:

$$\frac{\int_r^s (w \cdot f)}{\int_r^s w}.$$

A *w-partition of an interval [r,s]* is any finite sequence of numbers $a_1,\ldots,a_n$ such that

$$r = a_1 < a_2 < \cdots < a_n = s$$

and

$$m[a_1,a_2] = \cdots = m[a_{n-1},a_n].$$

A w-partition with exactly n terms is said to be *n-fold*.

Because our definition of $\underline{\underline{M}}_r^s f$ is based on arithmetic partitions and weighted arithmetic averages, we were surprised to discover that $\underline{\underline{M}}_r^s f$ is equal to the limit of the convergent sequence whose nth term is the unweighted arithmetic average

$$\frac{f(a_1) + \cdots + f(a_n)}{n},$$

where $a_1,\ldots,a_n$ is the n-fold w-partition of [r,s].[2]

The operator $\underline{\underline{M}}$ is additive, subtractive, and homogeneous; that is, if f and g are continuous on [r,s], then

$$\underline{\underline{M}}_r^s(f + g) = \underline{\underline{M}}_r^s f + \underline{\underline{M}}_r^s g,$$
$$\underline{\underline{M}}_r^s(f - g) = \underline{\underline{M}}_r^s f - \underline{\underline{M}}_r^s g,$$
$$\underline{\underline{M}}_r^s(c \cdot f) = c \cdot \underline{\underline{M}}_r^s f, \quad \text{c any constant.}$$

The operator $\underline{\underline{M}}$ is characterized by the following three properties.

For any interval [r,s] and any constant function $k(x) = c$ on [r,s],

$$\underline{\underline{M}}_r^s k = c.$$

For any interval [r,s] and any continuous functions f and g, if $f(x) \le g(x)$ on [r,s], then

$$\underline{\underline{M}}_r^s f \le \underline{\underline{M}}_r^s g.$$

For any numbers r, s, t such that $r < s < t$ and any continuous function f on [r,t],

$$m[r,s] \cdot \underline{\underline{M}}_r^s f + m[s,t] \cdot \underline{\underline{M}}_s^t f = m[r,t] \cdot \underline{\underline{M}}_r^t f.$$

Clearly, if w is constant on [r,s], then $\underline{\underline{M}}_r^s f$ equals $M_r^s f$.

We conclude this section with the following mean value theorem:

If f is continuous on [r,s], then strictly between r and s there is a number c such that

$$\underline{\underline{M}}_r^s f = f(c);$$

that is, the w-arithmetic-average of a continuous function on [r,s] is assumed at some argument strictly between r and s.

N O T E S

1. The w-arithmetic-average fits naturally into certain systems of calculus in a manner explained fully in [5] and [6].

2. A comparison of that result with the definition of $M_r^s f$ in Section 1.2 reveals that the method of partitioning argument intervals has a profound effect on the resulting average.

CHAPTER TWO

GEOMETRIC AVERAGES

2.1 INTRODUCTION

Next to the arithmetic averages, the geometric averages of functions are the best known and most often used. Geometric averages are relatively easy to calculate, and they may provide useful and intuitively-satisfying insight into certain scientific ideas. (For example, see page 54 of [3].)

In this chapter we shall discuss geometric averages in a manner that parallels our presentation of arithmetic averages in Chapter 1.

2.2 GEOMETRIC AVERAGE

Recall that our definition of the arithmetic average of a continuous function on an interval is based on arithmetic partitions and arithmetic averages (of n numbers). Our definition of the geometric average of a continuous positive function on an interval is based on arithmetic partitions and the following familiar concept.

The <u>geometric average of n positive numbers</u> $z_1,\ldots,z_n$ is the positive number $(z_1 \cdot z_2 \cdots z_n)^{1/n}$. (See Note 1.)

The <u>geometric average of a continuous positive function f on an interval [r,s]</u> is denoted by $\tilde{M}_r^s f$ and is defined to be the positive limit of the convergent sequence whose nth term is the geometric average

$$[f(a_1) \cdot f(a_2) \cdots f(a_n)]^{1/n},$$

where $a_1,\ldots,a_n$ is the n-fold arithmetic partition of $[r,s]$.[2] We set $\tilde{M}_s^r f = \tilde{M}_r^s f$.

It turns out that

$$\tilde{M}_r^s f = \exp\left\{M_r^s(\ln f)\right\} = \exp\left\{\frac{1}{s-r}\int_r^s (\ln f)\right\}.$$

The operator $\tilde{M}$ is multiplicative, divisional, and involutional; that is, if f and g are continuous and positive on $[r,s]$, then

$$\tilde{M}_r^s(f \cdot g) = \tilde{M}_r^s f \cdot \tilde{M}_r^s g,$$

$$\tilde{M}_r^s(f / g) = (\tilde{M}_r^s f) / (\tilde{M}_r^s g),$$

$$\tilde{M}_r^s(f^c) = (\tilde{M}_r^s f)^c, \quad \text{c any constant.}$$

The operator $\tilde{M}$ is characterized by the following three properties.

For any interval $[r,s]$ and any constant positive function $k(x) = c$ on $[r,s]$,

$$\tilde{M}_r^s k = c.$$

For any interval $[r,s]$ and any continuous positive functions f and g, if $f(x) \le g(x)$ on $[r,s]$, then

$$\tilde{M}_r^s f \le \tilde{M}_r^s g.$$

For any numbers r, s, t such that $r < s < t$ and any continuous positive function f on $[r,t]$,

$$(\tilde{M}_r^s f)^{s-r} \cdot (\tilde{M}_s^t f)^{t-s} = (\tilde{M}_r^t f)^{t-r}.$$

We conclude this section with the following mean value theorem:

> If f is continuous and positive on [r,s], then strictly between r and s there is a number c such that
>
> $$\tilde{M}_r^s f = f(c);$$
>
> that is, the geometric average of a continuous positive function on [r,s] is assumed at some argument strictly between r and s.

NOTES

1. The geometric average of $z_1, \ldots, z_n$ is equal to

$$\exp\left\{\frac{\ln z_1 + \cdots + \ln z_n}{n}\right\}.$$

2. The geometric average fits naturally into a certain non-Newtonian calculus, called the geometric calculus, in a manner explained fully in [3].

2.3 GEOMETRIC WEIGHT FUNCTIONS

Our discussion of weighted geometric averages in Section 2.4 will be facilitated by the following definition.

A geometric weight function is any continuous function w such that $w(x) > 1$ on R.[1] There are, of course, infinitely-many geometric weight functions.

For the remainder of this chapter w is an arbitrarily selected geometric weight function.[2]

NOTES

1. The reason for the requirement that $w(x) > 1$ on R is given in Section 4.3.

2. We omit the theory of "geometric measures," which corresponds to the theory of classical measures in Section 1.4. The omitted material can easily be obtained as a special case of the general theory presented in Section 4.4.

2.4 WEIGHTED GEOMETRIC AVERAGES

Recall that our definition of the geometric average of a continuous positive function on an interval is based on arithmetic partitions and unweighted geometric averages (of n positive numbers). Our definition of the w-geometric-average of a continuous positive function is based on arithmetic partitions and weighted geometric averages (of n positive numbers), which we define next.

Let $v_1,\ldots,v_n$ be any n numbers that are all greater than 1. For any n positive numbers $z_1,\ldots,z_n$, the positive number

$$\left[z_1^{\ln v_1} \cdot z_2^{\ln v_2} \cdots z_n^{\ln v_n}\right]^{1/(\ln v_1 + \cdots + \ln v_n)}$$

is called a <u>weighted geometric average</u> of $z_1,\ldots,z_n$.[1]

The _w-geometric-average of a continuous positive function f on an interval [r,s]_ is denoted by $\underset{\equiv}{\tilde{M}}_r^s f$ and is defined to be the positive limit of the convergent sequence whose nth term is the weighted geometric average

$$\left[f(a_1)^{\ln w(a_1)} \cdot f(a_2)^{\ln w(a_2)} \cdots f(a_n)^{\ln w(a_n)}\right]^{1/(\ln w(a_1) + \cdots + \ln w(a_n))},$$

where $a_1, \ldots, a_n$ is the n-fold arithmetic partition of [r,s].[2]

It turns out that $\underset{\equiv}{\tilde{M}}_r^s f$ equals the well-known weighted geometric average of f on [r,s]:

$$\exp\left[\frac{\int_r^s (\ln w \cdot \ln f)}{\int_r^s \ln w}\right].$$

Notice that the expression within the brackets represents a weighted arithmetic average of $\ln f$.

The operator $\underset{\equiv}{\tilde{M}}$ is multiplicative, divisional, and involutional; that is, if f and g are continuous and positive on [r,s], then

$$\underset{\equiv}{\tilde{M}}_r^s (f \cdot g) = \underset{\equiv}{\tilde{M}}_r^s f \cdot \underset{\equiv}{\tilde{M}}_r^s g,$$

$$\underset{\equiv}{\tilde{M}}_r^s (f / g) = (\underset{\equiv}{\tilde{M}}_r^s f) / (\underset{\equiv}{\tilde{M}}_r^s g),$$

$$\underset{\equiv}{\tilde{M}}_r^s (f^c) = (\underset{\equiv}{\tilde{M}}_r^s f)^c, \quad \text{c any constant.}$$

The operator $\underset{\equiv}{\tilde{M}}$ is characterized by the following three properties.

For any interval [r,s] and any constant positive function $k(x) = c$ on [r,s],

$$\underset{\equiv}{\tilde{M}}_r^s f = c.$$

For any interval [r,s] and any continuous positive

functions f and g, if $f(x) \le g(x)$ on [r,s], then

$$\tilde{\underline{\underline{M}}}_r^s f \le \tilde{\underline{\underline{M}}}_r^s g.$$

For any numbers r, s, t such that r < s < t and any continuous positive function f on [r,t],

$$\left[\tilde{\underline{\underline{M}}}_r^s f\right]^{\int_r^s \ln w} \cdot \left[\tilde{\underline{\underline{M}}}_s^t f\right]^{\int_s^t \ln w} = \left[\tilde{\underline{\underline{M}}}_r^t f\right]^{\int_r^t \ln w}.$$

Clearly, if w is constant on [r,s], then $\tilde{\underline{\underline{M}}}_r^s f$ equals $\tilde{M}_r^s f$.

We conclude this section with the following mean value theorem.

If f is continuous and positive on [r,s], then strictly between r and s there is a number c such that

$$\tilde{\underline{\underline{M}}}_r^s f = f(c);$$

that is, the w-geometric-average of a continuous positive function on [r,s] is assumed at some argument strictly between r and s.

N O T E S

1. That weighted geometric average is equal to

$$\exp\left\{\frac{\ln v_1 \cdot \ln z_1 + \cdots + \ln v_n \cdot \ln z_n}{\ln v_1 + \cdots + \ln v_n}\right\},$$

and reduces to the unweighted geometric average of $z_1,\ldots,z_n$ when $v_1 = \cdots = v_n = e$.

2. The w-geometric-average fits naturally into a certain system of calculus in a manner explained fully in [5].

CHAPTER THREE

SYSTEMS OF ARITHMETIC

3.1 INTRODUCTION

Arithmetic averages are based on classical arithmetic, which is usually called the real number system. But it was our use of nonclassical arithmetics that led to the general theory of averages to be presented in Chapter 4. The use of nonclassical arithmetics also led to the general theory of non-Newtonian calculus, to the development of non-Cartesian analytic geometries, to the creation of a new theory of subjective probability, and to the conception of new kinds of vectors, centroids, least-squares methods, and complex numbers.[1] Furthermore, nonclassical arithmetics may be useful for devising new systems of measurement that will yield simpler or new physical laws. This was clearly recognized by Norman Robert Campbell, a pioneer in the theory of measurement:

> "...we must recognize the possibility that a system of measurement may be arbitrary otherwise than in the choice of unit; there may be arbitrariness in the choice of the process of addition."[2]

In this chapter we discuss the general concept of an arithmetic and we present some specific arithmetics of special interest.

NOTES

1. See [2], [3], [4], [5], and [7].

The nonclassical arithmetics and non-Newtonian calculi should be distinguished from the nonstandard arithmetic and analysis developed by the logicians.

2. The quotation is from Campbell's remarkable book *Foundations of Science* (Dover reprint, 1957), p.292.

3.2 CLASSICAL ARITHMETIC

Classical arithmetic has been used for centuries but was not established on a sound axiomatic basis until the latter part of the nineteenth century. However, the details of such a treatment are not essential here.[1]

Informally, classical arithmetic (or the real number system) is a system consisting of a set R, for which there are four operations +, −, ×, / and an ordering relation <, all subject to certain familiar axioms. The members of R are called (real) numbers, and we call R the realm of classical arithmetic.

NOTE

1. A new axiomatic development of classical arithmetic and a novel approach to basic logic are presented in [1].

3.3 ARITHMETICS

The concept of a complete ordered field evolved from the axiomatization of classical arithmetic. Informally, a *complete ordered field* is a system consisting of a set A, and four operations $\dot{+}$, $\dot{-}$, $\dot{\times}$, $\dot{/}$ and an ordering relation $\dot{<}$, all of which behave with respect to A exactly as +, −, ×, /, < behave with respect to R. We call A the *realm of* $(A, \dot{+}, \dot{-}, \dot{\times}, \dot{/}, \dot{<})$.

By an *arithmetic* we mean a complete ordered field whose realm is a subset of R. There are infinitely-many arithmetics, one of which is classical arithmetic. Any two arithmetics are structurally equivalent (isomorphic).

The rules for handling any arithmetic $(A, \dot{+}, \dot{-}, \dot{\times}, \dot{/}, \dot{<})$ are exactly the same as the rules for handling classical arithmetic. For example, $\dot{+}$ and $\dot{\times}$ are commutative and associative; $\dot{\times}$ is distributive with respect to $\dot{+}$; and $\dot{<}$ is transitive.

3.4 α-ARITHMETIC

A *generator* is any one-to-one function whose domain is R and whose range is a subset of R. For example, I (the identity function) and exp are generators.

Consider any generator α and let A be the range of α. By *α-arithmetic* we mean the arithmetic whose realm is A and whose operations and ordering relation are defined for A as follows.

α-addition: $y \dot{+} z = \alpha(\alpha^{-1}(y) + \alpha^{-1}(z))$.

α-subtraction: $y \dot{-} z = \alpha(\alpha^{-1}(y) - \alpha^{-1}(z))$.

α-multiplication: $y \dot{\times} z = \alpha(\alpha^{-1}(y) \times \alpha^{-1}(z))$.

α-division: $y \dot{/} z = \alpha(\alpha^{-1}(y) / \alpha^{-1}(z))$ if $z \neq \alpha(0)$.

α-order: $y \dot{<} z$ if and only if $\alpha^{-1}(y) < \alpha^{-1}(z)$.

We say that α <u>generates</u> α-arithmetic. For example, the identity function I generates classical arithmetic and the function exp generates geometric arithmetic, which is discussed in Section 3.6. Each generator generates exactly one arithmetic, and, conversely, each arithmetic is generated by exactly one generator.

All concepts in classical arithmetic have natural counterparts in α-arithmetic, some of which we shall discuss shortly.

For each number x, we set $\dot{x} = \alpha(x)$.

Since

$$y \dot{+} \dot{0} = y \text{ and } y \dot{\times} \dot{1} = y$$

for each number y in A, we call $\dot{0}$ and $\dot{1}$ the <u>α-zero</u> and the <u>α-one</u>, respectively.

The <u>α-integers</u> are the numbers $\dot{n}$, where n is an arbitrary integer; if $\dot{0} \dot{<} \dot{n}$, then

$$\dot{n} = \underbrace{\dot{1} \dot{+} \cdots \dot{+} \dot{1}}_{n \text{ terms}}.$$

The <u>α-positive numbers</u> are the numbers x in A such that $\dot{0} \dot{<} x$. The <u>α-negative numbers</u> are the numbers x in A for which $x \mathbin{\dot{<}} \dot{0}$.

For each number y in A, we make the following definitions:

$$\dot{-}y = \dot{0} \mathbin{\dot{-}} y;$$

$$y^{\dot{2}} = y \mathbin{\dot{\times}} y; \quad \text{(See Note 1.)}$$

$$\dot{|}y\dot{|} = \begin{cases} y & \text{if } \dot{0} \mathbin{\dot{\le}} y \\ \dot{-}y & \text{if } y \mathbin{\dot{<}} \dot{0}; \quad \text{(See Note 2.)} \end{cases}$$

if $\dot{0} \mathbin{\dot{\le}} y$, then $\dot{\sqrt{y}}$ is the unique number z in A such that $\dot{0} \mathbin{\dot{\le}} z$ and $z^{\dot{2}} = y$.

It turns out that

$$\dot{-}(\dot{-}y) = y;$$

$$(\dot{\sqrt{y}})^{\dot{2}} = y \quad \text{if } \dot{0} \mathbin{\dot{\le}} y;$$

$$\dot{\sqrt{y^{\dot{2}}}} = \dot{|}y\dot{|}.$$

Also,

$$\dot{-}y = \alpha(-\alpha^{-1}(y));$$

$$y^{\dot{2}} = \alpha([\alpha^{-1}(y)]^2);$$

$$\dot{|}y\dot{|} = \alpha(|\alpha^{-1}(y)|);$$

$$\dot{\sqrt{y}} = \alpha(\sqrt{\alpha^{-1}(y)}) \quad \text{if } \dot{0} \mathbin{\dot{\le}} y.$$

An <u>α-progression</u> (or <u>natural progression in α-arithmetic</u>) is a finite sequence of numbers $y_1, \ldots, y_n$ in A such that

$$y_2 \mathbin{\dot{-}} y_1 = \cdots = y_n \mathbin{\dot{-}} y_{n-1}.$$

In classical arithmetic the natural progressions are the arithmetic progressions; in geometric arithmetic they are the geometric progressions.

For any numbers r and s in A, if $r \dot{<} s$, then the set consisting of all numbers x in A such that $r \dot{\leq} x \dot{\leq} s$ is called an <u>α-interval</u> and is denoted by $\dot{[}r,s\dot{]}$. The <u>α-interior of $\dot{[}r,s\dot{]}$</u> consists of all numbers x in A for which $r \dot{<} x \dot{<} s$.

An <u>α-partition of an α-interval $\dot{[}r,s\dot{]}$</u> is any α-progression whose first and last terms are r and s, respectively. An α-partition with exactly n terms is said to be <u>n-fold</u>.

Let $\{y_n\}$ be an infinite sequence of numbers in A. Then there is at most one number y in A that has the following property:

> For each α-positive number p,
> there is a positive integer m such that
> for any integer n, if $n \geq m$, then
> $\dot{|}y_n \dot{-} y\dot{|} \dot{<} p$.[3]

If there is such a number y, then $\{y_n\}$ is said to be <u>α-convergent to y</u> and y is called the <u>α-limit</u> of $\{y_n\}$.

Of course, if α = I, then α-convergence is identical with classical convergence.

Finally, it turns out that $\{y_n\}$ is α-convergent to y if and only if $\{\alpha^{-1}(y_n)\}$ is classically convergent to $\alpha^{-1}(y)$.

<u>N O T E S</u>

1. Since $\dot{2} = \alpha(2)$, there is a slight risk that the reader will take $y^{\dot{2}}$ to be $y^{\alpha(2)}$. We wish to stress that $y^{\dot{2}}$ is defined to be $y \dot{\times} y$, which equals $\alpha([\alpha^{-1}(y)]^2)$.

2. Of course, "$\dot{0} \mathrel{\dot{\le}} y$" is an abbreviation for "either $\dot{0} \mathrel{\dot{<}} y$ or $\dot{0} = y$."

3. It is helpful to think of $\dot{|}y_n \mathbin{\dot{-}} y\dot{|}$ as the "α-distance" from y_n to y.

3.5 α-AVERAGES

The <u>α-average of n numbers</u> $y_1, \ldots, y_n$ in A is the number

$$(y_1 \mathbin{\dot{+}} \cdots \mathbin{\dot{+}} y_n) \mathbin{\dot{/}} \dot{n},$$

which is in A and equals

$$\alpha([\alpha^{-1}(y_1) + \cdots + \alpha^{-1}(y_n)] / n).$$

Clearly, if $\alpha = I$, then the α-average is identical with the arithmetic average.

The following comparisons show that the role of the α-average in α-arithmetic is similar to the role of the arithmetic average in classical arithmetic.

Let z be the arithmetic average of n numbers $z_1, \ldots, z_n$, and let y be the α-average of n numbers $y_1, \ldots, y_n$ in A. Then z, which equals

$$(z_1 + \cdots + z_n) / n,$$

is the unique number such that

$$\underbrace{z + \cdots + z}_{n \text{ terms}} = z_1 + \cdots + z_n;$$

and y, which equals

$$(y_1 \mathbin{\dot{+}} \cdots \mathbin{\dot{+}} y_n) \mathbin{\dot{/}} \dot{n},$$

is the unique number in A such that

$$\underbrace{y \dot{+} \cdots \dot{+} y}_{n \text{ terms}} = y_1 \dot{+} \cdots \dot{+} y_n.$$

Thus, it is appropriate to say that the arithmetic average and the α-average are the natural averages of classical arithmetic and α-arithmetic, respectively.

Furthermore, for the arithmetic average z we have

(1) $(z - z_1) + \cdots + (z - z_n) = 0$;

(2) the expression

$$\sqrt{(x - z_1)^2 + \cdots + (x - z_n)^2}\,,$$

where x is unrestricted in R,

is a minimum when and only when $x = z$.

Similarly, for the α-average y we have

(3) $(y \dot{-} y_1) \dot{+} \cdots \dot{+} (y \dot{-} y_n) = \dot{0}$;

(4) the expression

$$\dot{\sqrt{(x \dot{-} y_1)^{\dot{2}} \dot{+} \cdots \dot{+} (x \dot{-} y_n)^{\dot{2}}}}\,,$$

where x is unrestricted in A,

has an α-minimum when and only when $x = y$.[1]

It is convenient to conceive the radical expression in (2) above as representing the "classical distance" from x to $z_1,\ldots,z_n$. (For $n = 1$, the expression reduces to $|x - z_1|$.) Accordingly, one may say that the arithmetic average of $z_1, \ldots, z_n$ is the number that is "classically closest" to $z_1, \ldots, z_n$.

Similarly, it is convenient to conceive the radical expression in (4) above as representing the "α-distance" from x to $y_1,\ldots,y_n$. (For n = 1, the expression reduces to $\dot{|}x \dot{-} y_1\dot{|}$.) Thus, one may say that the α-average of $y_1,\ldots,y_n$ is the number in A that is "α-closest" to $y_1,\ldots,y_n$.[2]

Our discussion of weighted averages parallels the preceding discussion of unweighted averages.

Let $w_1,\ldots,w_n$ be any n α-positive numbers. For any n numbers $y_1,\ldots,y_n$ in A, the number

$$\cdot\frac{(w_1 \dot{\times} y_1) \dot{+} \cdots \dot{+} (w_n \dot{\times} y_n)}{w_1 \dot{+} \cdots \dot{+} w_n}\cdot \qquad \text{(See Note 3.)}$$

is called a <u>weighted α-average</u> of $y_1,\ldots,y_n$.[4]

The following comparisons show that the role of weighted α-averages in α-arithmetic is similar to the role of weighted arithmetic averages in classical arithmetic.

Let $v_1,\ldots,v_n$ be any n positive numbers, and let z be the following weighted arithmetic average of n numbers $z_1, \ldots, z_n$:

$$\frac{v_1 z_1 + \cdots + v_n z_n}{v_1 + \cdots + v_n}.$$

Also, let $w_1,\ldots,w_n$ be any n α-positive numbers, and let y be the following weighted α-average of n numbers $y_1,\ldots,y_n$ in A:

$$\cdot\frac{(w_1 \dot{\times} y_1) \dot{+} \cdots \dot{+} (w_n \dot{\times} y_n)}{w_1 \dot{+} \cdots \dot{+} w_n}\cdot\ .$$

Then z is the unique number such that

$$v_1 z + \cdots + v_n z = v_1 z_1 + \cdots + v_n z_n;$$

and y is the unique number in A such that

$$(w_1 \dot{\times} y) \dot{+} \cdots \dot{+} (w_n \dot{\times} y) =$$
$$(w_1 \dot{\times} y_1) \dot{+} \cdots \dot{+} (w_n \dot{\times} y_n).$$

Furthermore, for the weighted arithmetic average z we have

(5) $v_1(z - z_1) + \cdots + v_n(z - z_n) = 0$;

(6) the expression

$$\sqrt{v_1(x - z_1)^2 + \cdots + v_n(x - z_n)^2}\,,$$

where x is unrestricted in R,

is a minimum when and only when x = z.

Similarly, for the weighted α-average y we have

(7) $[w_1 \dot{\times} (y \dot{-} y_1)] \dot{+} \cdots \dot{+} [w_n \dot{\times} (y \dot{-} y_n)] = \dot{0}$;

(8) the expression

$$\dot{\sqrt{[w_1 \dot{\times} (x \dot{-} y_1)^{\dot{2}}] \dot{+} \cdots \dot{+} [w_n \dot{\times} (x \dot{-} y_n)^{\dot{2}}]}}\,,$$

where x is unrestricted in A,

has an α-minimum when and only when x = y.

In view of (6) and (8) above, one may say that z is the number "classically closest" to $z_1,\ldots,z_n$ relative to $v_1,\ldots,v_n$, and that y is the number in A "α-closest" to $y_1,\ldots,y_n$ relative to $w_1,\ldots,w_n$.

N O T E S

1. Of course, the α-minimum (if there is one) of a set S of numbers in A is the unique number a in S such that for each number x in S, $a \dot{\leq} x$.

2. Although the α-average is widely known (in the case where α is continuous), we have seen no reference to it as the natural counterpart in α-arithmetic of the arithmetic average in classical arithmetic.

3. Of course, an expression such as $\cdot\frac{a}{b}\cdot$ means a /b.

4. That weighted α-average is in A, is equal to

$$\alpha\left\{\frac{\alpha^{-1}(w_1)\cdot\alpha^{-1}(y_1)+\cdots+\alpha^{-1}(w_n)\cdot\alpha^{-1}(y_n)}{\alpha^{-1}(w_1)+\cdots+\alpha^{-1}(w_n)}\right\},$$

and reduces to the unweighted α-average of $y_1,\ldots,y_n$ when $w_1 = \cdots = w_n = \dot{1}$.

3.6 GEOMETRIC ARITHMETIC

The arithmetic generated by the function exp will be called geometric arithmetic rather than exp-arithmetic. Similarly, the notions in geometric arithmetic will be indicated by the adjective "geometric" rather than the prefix "exp." For example, the natural average will be referred to as the geometric average, a usage that is consistent with generally accepted terminology as well as our terminology in Chapter 2.

Geometric arithmetic has the following features. (The letters y and z represent arbitrarily chosen positive numbers.)

Generator . . . . exp

Realm . . . . Set of all positive numbers

Geometric zero 1

Geometric one . . . . e

Geometric sum of y and z . . . . $\exp(\ln y + \ln z) = yz$

Geometric difference between y and z . . . . . . $\exp(\ln y - \ln z) = y/z$

Geometric product . . . . exp(ln y · ln z) = $y^{\ln z} = z^{\ln y}$

of y and z

Geometric quotient. . . . . exp(ln y / ln z) = $y^{1/\ln z}$

of y and z ($z \neq 1$)

Geometric order . . . . Identical with classical order

Geometric positive numbers . . Numbers greater than 1

Geometric negative numbers . . Positive numbers less than 1

Natural progressions . . . . . . Geometric progressions

Natural average . . . . Geometric average

Geometric convergence is equivalent to classical convergence in the sense that an infinite sequence $\{y_n\}$ of positive numbers geometrically converges to a positive number y if and only if $\{y_n\}$ classically converges to y.

Geometric arithmetic should be especially useful in situations where products and ratios provide the natural methods for combining and comparing magnitudes. Of course, geometric arithmetic applies only to positive numbers. However, a geometric-type arithmetic that applies to negative numbers can be obtained by using the generator -exp.

3.7 COMPARISON OF ARITHMETIC AND GEOMETRIC AVERAGES

By using geometric arithmetic we shall reveal some similarities between arithmetic averages and geometric averages. For the remainder of this section $\dot{+}$, $\dot{-}$, $\dot{\times}$, $\dot{/}$ denote the operations of geometric arithmetic and for each number x, we set

$\dot{x} = \exp x$.

The arithmetic average of n numbers $z_1,\ldots,z_n$ is the number $(z_1 + \cdots + z_n) / n$.

The geometric average of n positive numbers $z_1,\ldots,z_n$ is the positive number $(z_1 \dot{+} \cdots \dot{+} z_n) \dot{/} \dot{n}$.

If $v_1,\ldots,v_n$ are n positive numbers, then the number

$$\frac{v_1 z_1 + \cdots + v_n z_n}{v_1 + \cdots + v_n}$$

is a weighted arithmetic average of the n numbers z_1, $\ldots, z_n$.

If $v_1,\ldots,v_n$ are n geometrically-positive numbers, then the positive number

$$\dot{\frac{(v_1 \dot{\times} z_1) \dot{+} \cdots \dot{+} (v_n \dot{\times} z_n)}{v_1 \dot{+} \cdots \dot{+} v_n}}$$

is a weighted geometric average of the n positive numbers $z_1,\ldots,z_n$.

If f is continuous on [r,s] and $a_1,\ldots,a_n$ is the n-fold arithmetic partition of [r,s], then

$$M_r^s f = \lim_{n\to\infty} \left\{ [f(a_1) + \cdots + f(a_n)] / n \right\};$$

and, if furthermore f is positive, then

$$\tilde{M}_r^s f = \lim_{n\to\infty} \left\{ [f(a_1) \dot{+} \cdots \dot{+} f(a_n)] \dot{/} \dot{n} \right\}.$$

If f is continuous on $[r,s]$ and $a_1,\ldots,a_n$ is the n-fold arithmetic partition of $[r,s]$, and if w is a classical weight function, then

$$\underline{\underline{M}}_r^s f = \lim_{n\to\infty}\left\{\frac{w(a_1)\cdot f(a_1) + \cdots + w(a_n)\cdot f(a_n)}{w(a_1) + \cdots + w(a_n)}\right\};$$

and, if furthermore f is positive and w is a geometric weight function, then

$$\tilde{\underline{\underline{M}}}_r^s f =$$

$$\lim_{n\to\infty}\left\{\cdot\ \frac{[w(a_1) \dot{\times} f(a_1)] \dot{+} \cdots \dot{+} [w(a_n) \dot{\times} f(a_n)]}{w(a_1) \dot{+} \cdots \dot{+} w(a_n)}\ \cdot\right\}.$$

In Section 2.2, it was noted that the operator $\tilde{M}$ is multiplicative, divisional, and involutional, that is, if f and g are continuous and positive on $[r,s]$, then

$$\tilde{M}_r^s(f\cdot g) = \tilde{M}_r^s f \cdot \tilde{M}_r^s g,$$

$$\tilde{M}_r^s(f\,/\,g) = (\tilde{M}_r^s f)\,/\,(\tilde{M}_r^s g),$$

$$\tilde{M}_r^s(f^c) = (\tilde{M}_r^s f)^c, \quad c \text{ any constant.}$$

By using geometric arithmetic to re-express those three properties we find that they are actually conditions of additivity, subtractivity, and homogeneity within geometric arithmetic:

$$\tilde{M}_r^s(f \dot{+} g) = \tilde{M}_r^s f \dot{+} \tilde{M}_r^s g,$$

$$\tilde{M}_r^s(f \dot{-} g) = \tilde{M}_r^s f \dot{-} \tilde{M}_r^s g,$$

$$\tilde{M}_r^s(\dot{c} \dot{\times} f) = \dot{c} \dot{\times} \tilde{M}_r^s f, \quad c \text{ any constant.}$$

Of course, the corresponding properties of the operator $\tilde{\underline{\underline{M}}}$ may be re-expressed similarly.

3.8 POWER ARITHMETICS

In this section we show how to generate the infinitely-many power arithmetics, of which two particularly interesting arithmetics, called quadratic arithmetic and harmonic arithmetic, are special instances.

Let p be any nonzero number. The $\bar{p}$th-power function is the function that assigns to each number x the number

$$x^{\bar{p}} = \begin{cases} x^{p} & \text{if } 0 < x \\ 0 & \text{if } x = 0 \,. \\ -(-x)^{p} & \text{if } x < 0 \end{cases}$$

Notice that $x^{\bar{p}}$ is positive if x is positive, and negative if x is negative.

The $\bar{p}$th power function is one-to-one, is on R and onto R, has the $\overline{1/p}$th-power function as its inverse, and generates what we call pth-power arithmetic.

Some features of pth-power arithmetic are listed below. (The letters y and z represent arbitrarily chosen numbers.)

Generator . . . . $\bar{p}$th power function

Realm . . . . R

pth-Power zero 0

pth-Power one 1

pth-Power sum . . . . $(y^{\overline{1/p}} + z^{\overline{1/p}})^{\bar{p}}$

of y and z

pth-Power difference . . . . . . $(y^{\overline{1/p}} - z^{\overline{1/p}})^{\bar{p}}$

between y and z

pth-Power product . . . . $y \cdot z$

of y and z

pth-Power quotient . . . . y/z

of y and z ($z \neq 0$)

In pth-power arithmetic the natural average of n numbers $y_1, \ldots, y_n$ equals

$$[(y_1^{\overline{1/p}} + \cdots + y_n^{\overline{1/p}}) / n]^{\bar{p}},$$

which reduces to the well-known power average

$$[(y_1^{1/p} + \cdots + y_n^{1/p}) / n]^{p}$$

when $y_1, \ldots, y_n$ are all positive.[1]

Of course, if $p = 1$, pth-power arithmetic is identical with classical arithmetic.

By <u>quadratic arithmetic</u> we mean the pth-power arithmetic for which p equals 1/2. The quadratic sum of y and z reduces to the "Pythagorean sum"

$$\sqrt{y^2 + z^2}$$

when y and z are nonnegative, and in quadratic arithmetic the natural average of $y_1, \ldots, y_n$ reduces to the root mean square (or quadratic average)

$$\sqrt{[y_1^2 + \cdots + y_n^2] / n}$$

when $y_1, \ldots, y_n$ are all nonnegative.

By <u>harmonic arithmetic</u> we mean the pth-power arithmetic for which $p = -1$. The harmonic sum of y and z reduces to

$$1/(1/y + 1/z)$$

when y and z are positive, and in harmonic arithmetic the natural average of $y_1,\ldots,y_n$ reduces to the well-known harmonic average

$$\frac{1}{[1/y_1 + \cdots + 1/y_n] / n}$$

when $y_1,\ldots,y_n$ are all positive. Finally, in harmonic arithmetic the natural progressions of positive numbers are identical with the well-known harmonic progressions.

NOTE

1. The following result is well-known.

   If $a(p)$ is the pth-power average of n given positive numbers and if a is the geometric average of those numbers, then

$$\lim_{p\to\infty} a(p) = a.$$

3.9 SIGMOIDAL ARITHMETIC

Various arithmetics are generated by functions whose graphs are growth curves. Those curves include or are related to the logistic curves, cumulative normal curves, and the sigmoidal curves that occur in the study of population and biological growth.

For example, let $\sigma(x) = (e^x - 1)/(e^x + 1)$ on R. (Notice that $\sigma(2x) = \tanh x$.) Then the function σ is of the sigmoidal type (S-shaped graph) and generates sigmoidal arithmetic, whose realm consists of all numbers strictly between -1 and 1. The sigmoidal sum and sigmoidal difference of numbers y and z in the realm turn out to be $(y + z)/(1 + yz)$ and $(y - z)/(1 - yz)$, respectively.

It is entertaining to extend sigmoidal arithmetic by appending -1 and 1 to the realm and defining the extended sigmoidal sum of numbers y and z as follows:

$$y \dot{+} z = \begin{cases} (y + z)/(1 + yz) & \text{if } yz \neq -1 \\ 0 & \text{if } yz = -1 \end{cases} .$$

Then $1 \dot{+} (-1) = 0$; if $y \neq -1$, then $y \dot{+} 1 = 1$; and if $y \neq 1$, then $y \dot{+} (-1) = -1$. Thus, -1 and 1 behave as negative and positive infinity in extended sigmoidal arithmetic.

If units are chosen so that the speed of light is 1, then the extended sigmoidal sum of two velocities equals the relativistic composition of the velocities, even if one or both of the velocities is 1. (See Note 1.)

NOTE

1. We were interested to learn from [4] that any velocity composition rule consistent with the "Principle of Relativity" is expressible as an α-sum for some generator α.

CHAPTER FOUR

GENERAL THEORY OF AVERAGES OF FUNCTIONS

4.1 INTRODUCTION

For the remainder of this monograph, α and β are arbitrarily selected generators and $*$ ("star") is the ordered pair of arithmetics (α-arithmetic, β-arithmetic). The following notations will be useful.

	α-Arithmetic	β-Arithmetic
Realm	A	B
Addition	$\dot{+}$	$\ddot{+}$
Subtraction	$\dot{-}$	$\ddot{-}$
Multiplication . . .	$\dot{\times}$	$\ddot{\times}$
Division	$\dot{/}$ or ·—·	$\ddot{/}$ or ··—··
Order	$\dot{<}$	$\ddot{<}$

It should be understood that all definitions and results concerning α-arithmetic apply equally well to β-arithmetic. For example, the β-average of n numbers $y_1,\ldots,y_n$ in B is the following number in B: $(y_1 \ddot{+} \cdots \ddot{+} y_n) \ddot{/} \ddot{n}$. (Of course, $\ddot{n} = \beta(n)$.)

In our treatment of $*$-averages we shall apply α-arithmetic to function arguments and β-arithmetic to function values. Indeed, the $*$-averages apply only to functions with arguments in A and values in B. Accordingly, unless indicated or implied otherwise, all functions are assumed to be of that character.

For any function f defined at least on an α-interval containing the number a in its α-interior, we say that <u>f is ★-continuous at a</u> if and only if

for each β-positive number p,

there is an α-positive number q such that

for each number x in the domain of f,

if $\dot{|}x \dot{-} a\dot{|} \dot{<} q$, then $\ddot{|}f(x) \ddot{-} f(a)\ddot{|} \ddot{<} p$.[1]

When α and β are the identity function I, the concept of ★-continuity is identical with that of classical continuity, but that is possible even when α and β do not equal I.

The <u>isomorphism</u> ι (iota) from α-arithmetic to β-arithmetic is the unique function that possesses the following three properties.

1. ι is one-to-one.
2. ι is on A and onto B.
3. For any numbers u, v in A,

$$\iota(u \dot{+} v) = \iota(u) \ddot{+} \iota(v),$$

$$\iota(u \dot{-} v) = \iota(u) \ddot{-} \iota(v),$$

$$\iota(u \dot{\times} v) = \iota(u) \ddot{\times} \iota(v),$$

$$\iota(u \dot{/} v) = \iota(u) \ddot{/} \iota(v) \quad \text{if } v \neq \mathring{0},$$

$$u \dot{<} v \text{ if and only if } \iota(u) \ddot{<} \iota(v).$$

It turns out that for every number x in A, $\iota(x) = \beta(\alpha^{-1}(x))$. Also, for each number y, $\iota(\dot{y}) = \ddot{y}$.

Since, for example, $u \dot{+} v = \iota^{-1}(\iota(u) \ddot{+} \iota(v))$, it should be clear that any statement in α-arithmetic can readily be transformed into a statement in β-arithmetic.

<u>N O T E</u>

1. It is helpful to think of $\dot{|}x \mathbin{\dot{-}} a\dot{|}$ as the "α-distance" from x to a, and $\ddot{|}f(x) \mathbin{\ddot{-}} f(a)\ddot{|}$ as the "β-distance" from f(x) to f(a).

4.2 *-AVERAGE

Recall that our definition of the arithmetic average of a continuous function on an interval is based on arithmetic partitions and arithmetic averages (of n numbers). Also recall that our definition of the geometric average of a continuous positive function on an interval is based on arithmetic partitions and geometric averages (of n positive numbers). Our definition of the *-average of a *-continuous function on an α-interval is based on α-partitions and β-averages (of n numbers in B).

The <u>*-average of a *-continuous function f on an α-interval $\dot{[}r,s\dot{]}$</u> is denoted by $\overset{*}{M}{}_r^s f$ and is defined to be the β-limit of the β-convergent sequence whose nth term is the β-average

$$[f(a_1) \mathbin{\ddot{+}} \cdots \mathbin{\ddot{+}} f(a_n)] \mathbin{\ddot{/}} \ddot{n} ,$$

where $a_1,\ldots,a_n$ is the n-fold α-partition of $\dot{[}r,s\dot{]}$.[1] We set $\overset{*}{M}{}_s^r f = \overset{*}{M}{}_r^s f$.

The operator $\overset{*}{M}$ is β-additive, β-subtractive, and β-homogeneous; that is, if f and g are ∗-continuous on $[\dot{r},\dot{s}]$, then

$$\overset{*}{M}{}_r^s(f \ddot{+} g) = \overset{*}{M}{}_r^s f \ddot{+} \overset{*}{M}{}_r^s g,$$

$$\overset{*}{M}{}_r^s(f \ddot{-} g) = \overset{*}{M}{}_r^s f \ddot{-} \overset{*}{M}{}_r^s g,$$

$$\overset{*}{M}{}_r^s(c \ddot{\times} f) = c \ddot{\times} \overset{*}{M}{}_r^s f, \quad c \text{ any constant in B.}$$

The operator $\overset{*}{M}$ is characterized by the following three properties.

For any α-interval $[\dot{r},\dot{s}]$ and any constant function $k(x) = c$ on $[\dot{r},\dot{s}]$,

$$\overset{*}{M}{}_r^s k = c.$$

For any α-interval $[\dot{r},\dot{s}]$ and any ∗-continuous functions f and g, if $f(x) \ddot{\leq} g(x)$ on $[\dot{r},\dot{s}]$, then

$$\overset{*}{M}{}_r^s f \ddot{\leq} \overset{*}{M}{}_r^s g.$$

For any numbers r, s, t in A such that $r \dot{<} s \dot{<} t$ and any ∗-continuous function f on $[\dot{r},\dot{t}]$,

$$[\iota(s) \ddot{-} \iota(r)] \ddot{\times} \overset{*}{M}{}_r^s f \ddot{+} [\iota(t) \ddot{-} \iota(s)] \ddot{\times} \overset{*}{M}{}_s^t f$$
$$= [\iota(t) \ddot{-} \iota(r)] \ddot{\times} \overset{*}{M}{}_r^t f.$$

We conclude this section with the following mean value theorem:

If f is ∗-continuous on $[\dot{r},\dot{s}]$, then there is a number c in A such that $r \dot{<} c \dot{<} s$ and

$$\overset{*}{M}{}_r^s f = f(c);$$

thus, the ∗-average of a ∗-continuous function on $[\dot{r},\dot{s}]$ is assumed at some argument.

N O T E

1. The *-average fits naturally into a certain system of calculus in a manner explained fully in [2].

4.3 *-WEIGHT FUNCTIONS

Our discussion of weighted *-averages in Section 4.5 will be facilitated by the definitions and results in this section and the next.

A *-weight function is any *-continuous β-positive[1] function on A.[2] There are infinitely-many *-weight functions.

For the remainder of this chapter w is an arbitrarily selected *-weight function.

N O T E S

1. Of course, a β-positive function is a function whose values are all β-positive numbers.
2. In Section 2.3, we defined a geometric weight function to be any continuous function w such that $w(x) > 1$ on R. We require $w(x) > 1$ on R because we want the geometric weight functions to be the *-weight functions for which $\alpha = I$ and $\beta = \exp$, and in that case requiring that w be β-positive on A (in the definition of *-weight function) is equivalent to requiring $w(x) > 1$ on R.

4.4 *-MEASURES

The <u>*-measure of an α-interval $\dot{[}r,s\dot{]}$</u> is the β-positive number $\iota(s) \ddot{-} \iota(r)$.[1]

Now observe that the *-measure of $\dot{[}r,s\dot{]}$ equals the β-limit of the constant (and hence β-convergent) sequence whose nth term is the β-sum

$$\ddot{1} \ddot{\times} [\iota(a_2) \ddot{-} \iota(a_1)] \ddot{+} \cdots \ddot{+} \ddot{1} \ddot{\times} [\iota(a_n) \ddot{-} \iota(a_{n-1})],$$

where $a_1,\ldots,a_n$ is the n-fold α-partition of $\dot{[}r,s\dot{]}$. This suggests the following definition.

The <u>w*-measure of an α-interval $\dot{[}r,s\dot{]}$</u> is the β-positive limit of the β-convergent sequence whose nth term is the weighted β-sum

$$w(a_1) \ddot{\times} [\iota(a_2) \ddot{-} \iota(a_1)] \ddot{+} \cdots \ddot{+} w(a_{n-1}) \ddot{\times} [\iota(a_n) \ddot{-} \iota(a_{n-1})],$$

where $a_1,\ldots,a_n$ is the n-fold α-partition of $\dot{[}r,s\dot{]}$.

We shall use the symbol $\overset{*}{\dot{m}}\dot{[}r,s\dot{]}$ for the w*-measure of $\dot{[}r,s\dot{]}$. (In Section 4.6, we shall provide a formula for $\overset{*}{\dot{m}}\dot{[}r,s\dot{]}$.)

Of course, if $w(x) = \ddot{1}$ on $\dot{[}r,s\dot{]}$, then $\overset{*}{\dot{m}}\dot{[}r,s\dot{]}$ equals $\iota(s) \ddot{-} \iota(r)$, the *-measure of $\dot{[}r,s\dot{]}$.

Furthermore, if $r \dot{<} t \dot{<} s$, then

$$\overset{*}{\dot{m}}\dot{[}r,t\dot{]} \ddot{+} \overset{*}{\dot{m}}\dot{[}t,s\dot{]} = \overset{*}{\dot{m}}\dot{[}r,s\dot{]}.$$

<u>NOTE</u>

1. Although we could define the *-measure of $\dot{[}r,s\dot{]}$ to be the α-positive number $s \dot{-} r$, it is more convenient to use $\iota(s) \ddot{-} \iota(r)$, which equals $\iota(s \dot{-} r)$.

4.5 WEIGHTED *-AVERAGES

We remind the reader that w is an arbitrarily selected *-weight function.

Recall that our definition of the *-average of a *-continuous function on an α-interval is based on α-partitions and unweighted β-averages. Our definition of the w*-average of a *-continuous function on an α-interval is based on α-partitions and weighted β-averages.

The <u>w*-average of a *-continuous function f on an α-interval $\dot{[}r,s\dot{]}$</u> is denoted by $\overset{*}{\underline{\underline{M}}}{}_r^s f$ and is defined to be the β-limit of the β-convergent sequence whose nth term is the weighted β-average

$$\frac{[w(a_1)\ \ddot{\times}\ f(a_1)]\ \ddot{+} \cdots \ddot{+}\ [w(a_n)\ \ddot{\times}\ f(a_n)]}{w(a_1)\ \ddot{+} \cdots \ddot{+}\ w(a_n)},$$

where $a_1,\ldots,a_n$ is the n-fold α-partition of $\dot{[}r,s\dot{]}$.[1]

A <u>w*-partition of an α-interval $\dot{[}r,s\dot{]}$</u> is any finite sequence of numbers $a_1,\ldots,a_n$ in A such that

$$r = a_1\ \dot{<}\ a_2\ \dot{<} \cdots \dot{<}\ a_n = s$$

and

$$\overset{*}{m}\dot{[}a_1,a_2\dot{]} = \cdots = \overset{*}{m}\dot{[}a_{n-1},a_n\dot{]}.$$

A w*-partition with exactly n terms is said to be <u>n-fold</u>.

Because our definition of $\overset{*}{\underline{\underline{M}}}{}_r^s f$ is based on α-partitions and weighted β-averages, we were surprised to discover that $\overset{*}{\underline{\underline{M}}}{}_r^s f$ is equal to the β-limit of the β-convergent sequence whose nth term is the unweighted β-average

$$[f(a_1)\ \ddot{+} \cdots \ddot{+}\ f(a_n)]\ \ddot{/}\ \ddot{n},$$

where $a_1, \ldots, a_n$ is the n-fold $\omega*$-partition of $\dot{[}r,s\dot{]}$.

The operator $\overset{*}{\underline{\underline{M}}}$ is β-additive, β-subtractive, and β-homogeneous; that is, if f and g are *-continuous on $\dot{[}r,s\dot{]}$, then

$$\overset{*}{\underline{\underline{M}}}{}_r^s(f \ddot{+} g) = \overset{*}{\underline{\underline{M}}}{}_r^s f \ddot{+} \overset{*}{\underline{\underline{M}}}{}_r^s g,$$
$$\overset{*}{\underline{\underline{M}}}{}_r^s(f \ddot{-} g) = \overset{*}{\underline{\underline{M}}}{}_r^s f \ddot{-} \overset{*}{\underline{\underline{M}}}{}_r^s g,$$
$$\overset{*}{\underline{\underline{M}}}{}_r^s(c \ddot{\times} f) = c \ddot{\times} \overset{*}{\underline{\underline{M}}}{}_r^s f, \quad \text{c any constant in B.}$$

The operator $\overset{*}{\underline{\underline{M}}}$ is characterized by the following three properties.

For any α-interval $\dot{[}r,s\dot{]}$ and any constant function $k(x) = c$ on $\dot{[}r,s\dot{]}$,

$$\overset{*}{\underline{\underline{M}}}{}_r^s k = c.$$

For any α-interval $\dot{[}r,s\dot{]}$ and any *-continuous functions f and g, if $f(x) \ddot{\leq} g(x)$ on $\dot{[}r,s\dot{]}$, then

$$\overset{*}{\underline{\underline{M}}}{}_r^s f \ddot{\leq} \overset{*}{\underline{\underline{M}}}{}_r^s g.$$

For any numbers r, s, t in A such that $r \dot{<} s \dot{<} t$ and any *-continuous function f on $\dot{[}r,t\dot{]}$,

$$(\overset{*}{m}\dot{[}r,s\dot{]} \ddot{\times} \overset{*}{\underline{\underline{M}}}{}_r^s f) \ddot{+} (\overset{*}{m}\dot{[}s,t\dot{]} \ddot{\times} \overset{*}{\underline{\underline{M}}}{}_s^t f)$$
$$= (\overset{*}{m}\dot{[}r,t\dot{]} \ddot{\times} \overset{*}{\underline{\underline{M}}}{}_r^t f).$$

It turns out that if ω is constant on $\dot{[}r,s\dot{]}$, then $\overset{*}{\underline{\underline{M}}}{}_r^s f$ equals $\overset{*}{M}{}_r^s f$.

We conclude this section with the following mean value theorem:

If f is *-continuous on $\dot{[}r,s\dot{]}$, then there is a number c in A such that $r \dot{<} c \dot{<} s$ and

$$\overset{*}{\underline{\underline{M}}}{}_r^s f = f(c);$$

thus, the ω*-average of a *-continuous function on $[\dot{r},\dot{s}]$ is assumed at some argument.

N O T E

1. The ω*-average fits naturally into a certain system of calculus in a manner explained fully in [5].

4.6 RELATIONSHIP BETWEEN *-AVERAGES AND ARITHMETIC AVERAGES

In this section we shall indicate, among other things, the uniform relationship between the *-averages and the arithmetic averages.

For each number x in A, set $\bar{x} = \alpha^{-1}(x)$. And for each function f with arguments in A and values in B, set $\bar{f}(t) = \beta^{-1}(f(\alpha(t)))$.

Then f is *-continuous at a if and only if $\bar{f}$ is classically continuous at $\bar{a}$. (See Note 1.)

It can be shown that

$$(1)\quad \overset{*}{M}{}_r^s f = \beta\left\{M_{\bar{r}}^{\bar{s}}\bar{f}\right\} = \beta\left\{\frac{1}{\bar{s}-\bar{r}}\int_{\bar{r}}^{\bar{s}}\bar{f}\right\}.$$

Furthermore, it turns out that $\bar{\omega}$ is a classical weight function,

$$\overset{*}{m}[\dot{r},\dot{s}] = \beta(m[\bar{r},\bar{s}]) = \beta\left(\int_{\bar{r}}^{\bar{s}}\bar{\omega}\right),$$

and

$$(2)\quad \overset{*}{\underline{\underline{M}}}{}_r^s f = \beta\left\{\underline{\underline{M}}_{\bar{r}}^{\bar{s}}\bar{f}\right\} = \beta\left\{\frac{\int_{\bar{r}}^{\bar{s}}\bar{\omega}\,\bar{f}}{\int_{\bar{r}}^{\bar{s}}\bar{\omega}}\right\}.$$

(In (1) and (2), we assume f is $*$-continuous on $[\dot{r},\dot{s}]$.)

By letting $\alpha = I$ in (1) and (2), we obtain

$$(3)\quad \overset{*}{M}{}_r^s f = \beta\left\{\frac{1}{s - r}\int_r^s \beta^{-1}(f)\right\}$$

and

$$(4)\quad \underline{\underline{\overset{*}{M}}}{}_r^s f = \beta\left\{\frac{\int_r^s [\beta^{-1}(w)\cdot\beta^{-1}(f)]}{\int_r^s \beta^{-1}(w)}\right\}.$$

Expressions such as those on the right-hand sides of (3) and (4), accompanied by the qualification that β is classically continuous, appear occasionally in the literature; and it has long been recognized that such expressions represent averages.[2]

NOTES

1. Furthermore, if α and β are classically continuous at $\alpha^{-1}(a)$ and $\beta^{-1}(f(a))$, respectively, and if α^{-1} and β^{-1} are classically continuous at a and f(a), respectively, then f is $*$-continuous at a if and only if f is classically continuous at a.

2. For example, see *Inequalities* by Hardy, Littlewood, and Pólya (Cambridge University Press, 1952).

4.7 EXAMPLES

In this section we shall restrict our remarks to unweighted averages. However, similar remarks apply to weighted averages.

Each choice of specific generators for α and β determines a *-average. The following table indicates some of the infinitely-many possible choices. (We shall use the symbol g_p to denote the $\bar{p}$th-power function, which is defined in Section 3.8.)

*-average	α	β
arithmetic	I	I
geometric	I	exp
anageometric	exp	I
bigeometric	exp	exp
pth-power	I	g_p
ana-pth-power	g_p	I
bi-pth-power	g_p	g_p

The anageometric average is discussed in [2]. The bigeometric average is discussed in [2] and [7].

The pth-power average of a continuous positive function f on [r,s] is equal to the well-known power average

$$\left(\frac{1}{s - r} \int_r^s f^{1/p}\right)^p ,$$

which, if p = -1, reduces to the harmonic average.

In [2], the pth-power average, the ana-pth-power average, and the bi-pth-power average are discussed briefly in the cases where $p = 1/2$ and $p = -1$.

CHAPTER FIVE

HEURISTIC PRINCIPLES OF APPLICATION

5.1 INTRODUCTION

This chapter contains some heuristic principles that may be helpful for selecting appropriate averages.

Of course, one is always free to use any average that is meaningful in a given context. However, a suitable choice of an average depends chiefly upon its intended use.

We are not concerned here with probabilistic justifications for the use of a particular average. Nevertheless, we believe that any such justification for the use of arithmetic averages of numbers relative to classical arithmetic can be matched by a similar justification for the use of α-averages relative to α-arithmetic.

5.2 CHOOSING AVERAGES

Most of this section is devoted to α-averages of numbers rather than $*$-averages of functions, since the choice of the latter depends on how one would average function values. (Of course, the method of partitioning argument intervals is also a factor.)

Historically one reason for the popularity of the arithmetic average is its simplicity of calculation, but that issue is surely irrelevant in this age of computers. For instance, one investment advisory service had for many years maintained a geometric average of 1000 stocks.[1]

In choosing a method of averaging physical magnitudes one fundamental issue is the natural method of combining them. Where magnitudes are naturally combined by taking sums (unweighted or weighted), the (unweighted or weighted) arithmetic average is meaningful and may be useful. By the same token, where magnitudes are naturally combined by taking α-sums (unweighted or weighted), the (unweighted or weighted) α-average is meaningful and may be useful.

For example, where positive magnitudes are naturally combined by taking products (that is, geometric sums), the geometric average is meaningful and may be useful. And, as another example, since the total resistance of n electrical resistors connected in parallel equals the harmonic sum of the individual resistances, their harmonic average is meaningful and should be useful.

We shall let My_i represent the arithmetic average of n numbers $y_1,\ldots,y_n$, and if those numbers are all in A, we shall let $\dot{M}y_i$ represent their α-average.

Since the α-average and arithmetic average are the "natural" averages of α-arithmetic and classical arithmetic respectively (Section 3.5), the α-average has the same properties relative to α-arithmetic as the arithmetic average has relative to classical arithmetic. For example, since

$$(1) \quad M(y_i - z_i) = My_i - Mz_i,$$

one should expect that

$$(2) \quad \dot{M}(y_i \dot{-} z_i) = \dot{M}y_i \dot{-} \dot{M}z_i.$$

And indeed, the preceding equation is true.

Items (1) and (2) above are the basis of the heuristic principle that differences may best be averaged by the arithmetic average, but that α-differences may best be averaged by the α-average. For example, according to that principle, raios (of positive numbers) may best be averaged by the geometric average. However, there are situations where the arithmetic average of ratios is significant.[2]

For the remainder of this section we shall let $\tilde{M}u_i$ represent the geometric average of n positive numbers $u_1,\ldots,u_n$.

Consider the problem of estimating the area of a rectangle from n measurements $x_1,\ldots,x_n$ of its length and n measurements $y_1,\ldots,y_n$ of its width. The following four estimates will be considered.

I. $E_1 = (Mx_i)\cdot(My_i)$

II. $E_2 = M(x_i \cdot y_i)$

III. $E_3 = (\tilde{M}x_i)\cdot(\tilde{M}y_i)$

IV. $E_4 = \tilde{M}(x_i \cdot y_i)$

Because $E_1 \neq E_2$ except in isolated cases, and because $E_3 = E_4$, it would appear that here the geometric average is more appropriate than the arithmetic average. But that is to be expected since the geometric average is multiplicative and the area is the result of multiplication. (However, a similar analysis would indicate that the arithmetic average would be more appropriate than the geometric average if one were estimating the perimeter of the rectangle.) Furthermore, Method I, which is quite popular with scientists, has another disconcerting feature: If there arose new measurements x_{n+1} and y_{n+1} such that $x_{n+1} \cdot y_{n+1} = E_1$, then Method I applied to $x_1, \ldots, x_n, x_{n+1}$ and $y_1, \ldots, y_n, y_{n+1}$ *does not* yield the original estimate E_1 except in trivial cases. On the other hand, if there arose new measurements x_{n+1} and y_{n+1} such that $x_{n+1} \cdot y_{n+1} = E_3 = E_4$, then Methods III and IV applied to $x_1, \ldots, x_n, x_{n+1}$ and $y_1, \ldots, y_n, y_{n+1}$ *do* yield the original estimate E_3 (= E_4).

Now consider the problem of estimating the density of an object, given n measurements of its mass $u_1, \ldots, u_n$ and n measurements of its volume $v_1, \ldots, v_n$. There are at least four estimates of the density worthy of consideration.

$$E_1 = (Mu_i) / (Mv_i)$$
$$E_2 = M(u_i / v_i)$$
$$E_3 = (\tilde{M}u_i) / (\tilde{M}v_i)$$
$$E_4 = \tilde{M}(u_i / v_i)$$

Since $E_1 \neq E_2$ except in isolated cases, and since $E_3 = E_4$, it seems that the best choice ought to be the geometric average.

Some scientists, notably the psychophysicist S. S. Stevens of Harvard University, favor the use of certain invariance principles for choosing averages.[3]

Finally, in many situations where integrals are used, averages can be used instead to provide intuitively more satisfying results. Consider, for instance, a particle moving rectilinearly with positive velocity v. The distance s traveled in the time interval [a,b] is given by

$$s = \int_a^b v \, .$$

Although that fact may be clear to a student, he may nevertheless find that the following formula conveys a more immediate meaning:

$$s = (b - a) \cdot M_a^b v;$$

that is, the distance traveled equals the product of the time elapsed and the arithmetic average of the velocity. This version is a direct extension of the case where v is constant. Other examples will readily occur to the reader. (An example that involves the geometric average is indicated on page 54 of [3].)

NOTES

1. American Investors Service (Greenwich, Connecticut) distributed an interesting booklet by George A. Chestnut, Jr., who gave some excellent reasons why he considered geometric averaging the best method of averaging stock prices.

2. For example, suppose that initially $1000 is invested in one stock at $10 per share and $1000 in another stock at $20 per share. Subsequently the stocks are worth $5 and $50 per share respectively. Since the original investment of $2000 increased in value to $3000, the overall ratio change in value

is 1.5, which equals the *arithmetic* average of the ratio changes, 0.5 and 2.5, for the individual stocks.

3. A detailed discussion is given by S. S. Stevens in his article "On the Averaging of Data," which appeared in Science, Volume 121 (January 28, 1955), pp.113-6. Some comments on Stevens' ideas may be found in Brian Ellis' book, *Basic Concepts of Measurement* (Cambridge University Press, 1966).

CHAPTER SIX

MEANS OF TWO POSITIVE NUMBERS

6.1 INTRODUCTION

In this last chapter we shall use various *-averages of the identity function to construct an infinite family of means of two positive numbers. Apparently, some of those means have until now been obtained only by taking limits of other means.[1]

Henceforth, x and y are distinct positive numbers. Recall that I stands for the identity function.

NOTE

1. Discussions of various means of numbers and their application to the theory of inequalities can be found in the following references.

Beckenbach, E.F. and Bellman, R. *Inequalities*. Berlin: Springer-Verlag, 1961.

Hardy, G. H., Littlewood, J.E., and Pólya, G. *Inequalities*. Cambridge: Cambridge University Press, 1952.

Leach, E. B. and Sholander, M.C. "Extended Mean Values." *The American Mathematical Monthly* (February 1978).

Stolarsky, K. B. "The Power and Generalized Logarithmic Means." *The American Mathematical Monthly* (August 1980).

Tang, J. "On the Construction and Interpretation of Means." *International Journal of Mathematical Education in Science and Technology* (forthcoming).

6.2 EXAMPLES

The following easily verified results are special cases of the general theory to be presented in the next section.

The arithmetic average of I from x to y (that is, $M_x^y I$) is equal to $(x + y)/2$.

The bigeometric average (see Section 4.7) of I from x to y is equal to $\sqrt{xy}$.

The anageometric average (see Section 4.7) of I from x to y is equal to $(x - y)/(\ln x - \ln y)$, which is called the logarithmic mean of x and y.[1]

Finally, the geometric average of I from x to y is equal to $e^{-1}(x^x/y^y)^{1/(x-y)}$, which is called the identric mean of x and y.[2]

The preceding observations suggest that at least some of the various *-averages of I from x to y are in fact equal to means of x and y.

NOTES

1. The logarithmic mean is discussed in the article by Leach and Sholander indicated at the note in the preceding section.
2. The identric mean is discussed in the same article by Leach and Sholander.

6.3 *-MEAN

Let α and β be generators such that α^{-1} and β^{-1} are classically continuous at each positive number. Set $\overset{*}{M}(x,y) = \overset{*}{M}{}_x^y I$.

Because it turns out that

(1) $\overset{*}{M}(x,y) = \overset{*}{M}(y,x)$ and

(2) $\overset{*}{M}(x,y)$ is strictly between x and y,

we may call $\overset{*}{M}(x,y)$ the <u>*-mean of x and y</u>.

By making specific choices for α and β, one can produce a wide variety of *-means of x and y. Some examples are given in the next section.

It follows from Section 4.6 that $\overset{*}{M}(x,y)$ is equal to

$$\beta\left\{\frac{1}{\bar{y}-\bar{x}}\int_{\bar{x}}^{\bar{y}}(\beta^{-1}(\alpha))\right\},$$

where $\bar{x} = \alpha^{-1}(x)$ and $\bar{y} = \alpha^{-1}(y)$.

It is interesting to note that if $\alpha = \beta$, then $\overset{*}{M}(x,y)$ is equal to

$$\alpha([\alpha^{-1}(x) + \alpha^{-1}(y)]/2),$$

which, of course, equals the α-average of x and y.

6.4 SPECIAL CASES

In the table on page 55 we list a few of the various *-means of x and y that can be generated by specific choices of α and β. Some of the indicated means are discussed in the references presented at the note in Section 6.1.

For each nonzero number p, we let h_p be the function that assigns to each number x the number

$$\begin{cases} x^{1/p} & \text{if } 0 < x \\ 0 & \text{if } x = 0 \\ -(-x)^{1/p} & \text{if } x < 0 \end{cases} .$$

Notice that h_p, which is a generator, is the inverse of the $\bar{p}$th-power function discussed in Section 3.8.

α	β	$\overset{*}{M}(x,y)$
I	I	$\dfrac{x + y}{2}$
exp	exp	$\sqrt{x\,y}$
h_p	h_p	$\left[\dfrac{x^p + y^p}{2}\right]^{1/p}$
exp	I	$\dfrac{x - y}{\ln x - \ln y}$
exp	h_p	$\left[\dfrac{x^p - y^p}{p(\ln x - \ln y)}\right]^{1/p}$
I	exp	$e^{-1}\left[\dfrac{x^x}{y^y}\right]^{1/(x-y)}$
h_p	exp	$e^{-1/p}\exp\left[\dfrac{x^p \ln x - y^p \ln y}{x^p - y^p}\right]$
I	h_p $(p\neq-1)$	$\left[\dfrac{x^{p+1} - y^{p+1}}{(p + 1)(x - y)}\right]^{1/p}$
h_p	h_{s-p} $(s\neq0)$	$\left[\dfrac{x^s - y^s}{x^p - y^p}\cdot\dfrac{p}{s}\right]^{1/(s-p)}$

h_p $(p \neq -1)$	I	$\frac{p}{p+1} \cdot \frac{x^{p+1} - y^{p+1}}{x^p - y^p}$
h_p $(p = -1)$	I	$xy \frac{\ln x - \ln y}{x - y}$

6.5 CONJECTURES

We have thus far been unsuccessful in finding a natural way to extend the *-means of two positive numbers to *-means of n numbers. If this could be done, then one might be able to define a new class of averages of functions by employing the same technique that was used for defining *-averages of functions in Section 4.2. And then, it might be possible to use these new averages of functions to define new integrals, which would then lead to new systems of calculus in which these averages play a natural role.

B I B L I O G R A P H Y

1. Katz, R. *Axiomatic Analysis*. Rockport, MA: Mathco, 1964. This textbook, which was prepared under the general editorship of Professor David V. Widder, contains an original approach to basic logic and a novel axiomatic treatment of the real number system.

2. Grossman, M. and Katz, R. *Non-Newtonian Calculus*. Rockport, MA: Mathco, 1972.
Included in this book, which was the first publication on non-Newtonian calculus, are discussions of nine specific non-Newtonian calculi, the general theory of non-Newtonian calculus, and heuristic guides for the application thereof.

3. Grossman, M. *The First Nonlinear System of Differential and Integral Calculus*. Rockport, MA: Mathco, 1979.
This book contains a detailed account of the geometric calculus, which was the first of the non-Newtonian calculi. Also included are discussions of the analogy that led to the discovery of that calculus, and some heuristic guides for its application.

4. Meginniss, J. R. "Non-Newtonian Calculus Applied to Probability, Utility, and Bayesian Analysis." *Proceedings of the American Statistical Association*: Business and Economics Statistics Section (1980), pp. 405-410.
This paper presents a new theory of probability suitable for the analysis of human behavior and decision making. The theory is based on the idea that subjective probability is governed by the laws of a non-Newtonian calculus and one of its corresponding arithmetics.

5. Grossman, J., Grossman, M., and Katz, R. *The First Systems of Weighted Differential and Integral Calculus*. Rockport, MA: Archimedes Foundation, 1980.
This monograph reveals how weighted averages, Stieltjes integrals, and derivatives of one function with respect to another can be linked to form systems of calculus, which are called weighted calculi because in each such system a weight function plays a central role.

6. Grossman, J. *Meta-Calculus: Differential and Integral*. Rockport, MA: Archimedes Foundation, 1981.
This monograph contains a development of systems of calculus, called meta-calculi, that transcend the classical calculus, for example in the following manner. In each meta-calculus the gradient, or average rate of change, of a function f on an interval [r,s] depends on ALL the points (x,f(x)) for which $r \leq x \leq s$, whereas the classical gradient $[f(s) - f(r)]/(s-r)$ depends only on the endpoints (r,f(r)) and (s,f(s)). The meta-calculi arose from the problem of measuring stock-price

performance when taking all intermediate prices into account.

7. Grossman, M. *Bigeometric Calculus: A System with a Scale-Free Derivative*. Rockport, MA: Archimedes Foundation, 1983. This book contains a detailed treatment of the bigeometric calculus, which has a derivative that is scale-free, i.e., invariant under all changes of scales (or units) in function arguments and values. Also included are heuristic guides for the application of that calculus, and various related matters such as the bigeometric method of least squares.

LIST OF SYMBOLS

I N D E X

427593

Made in the USA